Types of Spaceships: Past, Present, and Future

Copyright Page

This book is copyrighted for 2020

Types of Spaceships: Past, Present, and Future

The Living in Space Series Book 7

By Martin K. Ettington

ISBN: 9798683382407

Printed in the United States of America

Types of Spaceships: Past, Present, and Future

Types of Spaceships: Past, Present, and Future

This book is the seventh volume of "The Living in Space Series". Each volume focuses on one particular technology of living in space.

In this book the focus is on spaceships used to access and travel in space. In my research I was pretty amazed at how many spaceships have been designed, the large number available in the current era, and that many more are already on the drawing boards.

Some spaceships like the Space Shuttle seemed like a great idea back in the 1970s and 1980s. But due to the compromises in design were very expensive for each flight and kind of dangerous too. Even though the Space Shuttle was flown for over 30 years it was eventually cancelled because of cost issues.

The newer transportation systems focus on lowering costs through reusability. This will lead to a much better long term paradigm for expanding people's entry into space because of the lower costs reusability provides.

I also needed to determine what types of ships to include in this book. What is a valid spaceship? My decision was to use the Von Karmen Line which says that a spaceship flown higher than 50 miles (80 km) has entered space.

Using the Von Karmen Line includes one manned device—the X-15 experimental rocket plane as a spaceshlp although most people wouldn't think so.

Lastly, I've included some notional spaceship designs which might be used to travel to Mars or other bodies in the Solar System.

Types of Spaceships: Past, Present, and Future

Types of Spaceships: Past, Present, and Future

Other books by Martin K. Ettington

Types of Spaceships: Past, Present, and Future

How to Survive Anything: From the Wilderness to Man Made Disasters
Building and Stocking a Nuclear Shelter for less than $10,000
All About Mars Journeys and Settlement
Mining the Asteroid Belt

Ancient History
The Real Atlantis-In the Eye of the Sahara
Ancient & Prehistoric Civilizations
Ancient & Prehistoric Civilizations-Book Two
The History of Antediluvian Giants
The Antediluvian History of Earth
Ancient Underground Cities and Tunnels
Strange Objects Which Should Not Exist

Strange and Ancient Places in the USA
A Theory of Ancient Prehistory And Giant Aliens
Aliens and Space
Aliens and Secret Technology
Aliens Are Already Among Us
Designing and Building Space Colonies
Humanity and the Universe
All About Moon Bases
All About Mars Journeys and Settlement
The Space and Aliens Six Books Bundle
A Theory of Ancient Prehistory and Giant Aliens
The Space Colonies and Space Structures Coloring Book
All About Asteroids

The Longevity Training Series

(A transcription of the online Multimedia Longevity Coaching Training Program)

The Personal Longevity Training Series-Book1-Long Lived Persons
The Personal Longevity Training Series-Book2-Your Soul's Purpose
The Personal Longevity Training Series-Book3-Enable Your Life Urge
The Personal Longevity Training Series-Book4-Your Spiritual Connection
The Personal Longevity Training Series-Book5-Having Love in Your Heart
The Personal Longevity Training Series-Book6-Energy Body Health
The Personal Longevity Training Series-Book7-The Science of Longevity
The Personal Longevity Training Series-Book8-Physical Body Health
The Personal Longevity Training Series-Book9-Avoiding Accidents
The Personal Longevity Training Series-Book10-Implementing These Principles

The Personal Longevity Training Series-Books One Thru Ten

These books are all available in digital and printed formats from my website and on Amazon, Barnes & Noble, Apple ITunes, and many other sites

My Books Website is: http://mkettingtonbooks.com

Types of Spaceships: Past, Present, and Future

<u>Signup for our Mailing List to get the following:</u>

1) A discount coupon for 25% discount on all books on our site

2) Occasional Notices of new books available

3) Occasional Email on other offerings of ours (Monthly)

Go to this link to sign-up:

http://personal-longevity.com/mkebooks/emailsignup/

And click this link to get the FREE 102 page Ebook titled "Secrets of Many Things"

If you have any questions about this book or other subjects please contact the Author at:

mke@mkettingtonbooks.com

Types of Spaceships: Past, Present, and Future

Types of Spaceships: Past, Present, and Future

Table of Contents

Types of Spaceships: Past, Present, and Future

Types of Spaceships: Past, Present, and Future

1.0 Introduction

This book is the seventh volume of "The Living in Space Series". Each volume focuses on one particular technology of living in space.

In this book the focus is on spaceships used to access and travel in space. In my research I was pretty amazed at how many spaceships have been designed, the large number available in the current era, and that many more are already on the drawing boards.

Some spaceships like the Space Shuttle seemed like a great idea back in the 1970s and 1980s. But due to the compromises in design were very expensive for each flight and kind of dangerous too. Even though the Space Shuttle was flown for over 30 years it was eventually cancelled because of cost issues.

The newer transportation systems focus on lowering costs through reusability. This will lead to a much better long term paradigm for expanding people's entry into space because of the lower costs reusability provides.

I also needed to determine what types of ships to include in this book. What is a valid spaceship? My decision was to use the Von Karmen Line which says that a spaceship flown higher than 50 miles (80 km) has entered space.

Using the Von Karmen Line includes one manned device—the X-15 experimental rocket plane as a spaceship although most people wouldn't think so.

Lastly, I've included some notional spaceship designs which might be used to travel to Mars or other bodies in the Solar System.

Types of Spaceships: Past, Present, and Future

Types of Spaceships: Past, Present, and Future

2.0 Past Spaceships

The earliest spaceship was actually the X-15 rocket plane first flown in 1959. It crossed the Von Karmen line to make the pilot the first man in space.

There are also many past spaceships in this chapter which helped man to reach the moon and build the space station.

2.1 USA-X-15 Rocketplane

Three veteran NASA X-15 pilots (John B. McKay, William H. Dana and Joseph Albert Walker) were retroactively (two posthumously) awarded their astronaut wings, as they had flown between 90 km (56 miles) and 108 km (67 miles) in the 1960s, but at the time had not been recognized as astronauts.

The North American X-15 was a hypersonic rocket-powered aircraft operated by the United States Air Force and the National Aeronautics and Space Administration as part of the X-plane series of experimental aircraft. The X-15 set speed and altitude records in the 1960s, reaching the edge of outer space and returning with valuable data used in aircraft and spacecraft design. The X-15's official world record for the highest speed ever recorded by a crewed, powered aircraft, set in October 1967 when William J. Knight flew at Mach 6.70 at 102,100 feet (31,120 m), a speed of 4,520 miles per hour (7,274 km/h; 2,021 m/s), has remained unbroken as of 2020

During the X-15 program, 12 pilots flew a combined 199 flights. Of these, 8 pilots flew a combined 13 flights which met the Air Force spaceflight criterion by exceeding the altitude of 50 miles (80 km), thus qualifying these pilots as being astronauts. The Air Force pilots qualified for military astronaut wings immediately, while the civilian pilots were eventually awarded NASA astronaut wings in 2005, 35 years after the last X-15 flight.

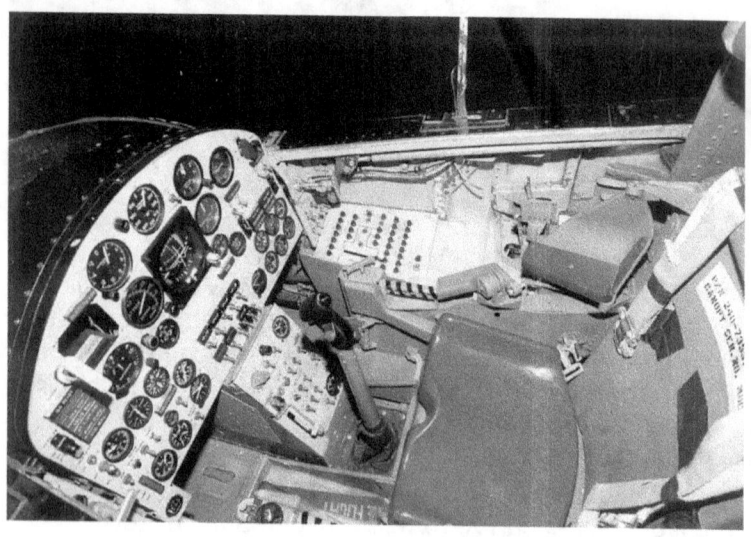

The X-15 Cockpit

Types of Spaceships: Past, Present, and Future

The first X-15 flight was an unpowered glide flight by Scott Crossfield, on 8 June 1959. Crossfield also piloted the first powered flight on 17 September 1959, and his first flight with the XLR-99 rocket engine on 15 November 1960. Twelve test pilots flew the X-15. Among these were Neil Armstrong, later a NASA astronaut and first man to set foot on the Moon, and Joe Engle, later a commander of NASA Space Shuttle missions.

Types of Spaceships: Past, Present, and Future

2.2 Soviet-Vostok

The Vostok was a type of spacecraft built by the Soviet Union. The first human spaceflight was accomplished with Vostok 1 on April 12, 1961, by Soviet cosmonaut Yuri Gagarin.

The spacecraft was part of the Vostok program, in which six crewed spaceflights were made, from 1961–1963. Two further human space flights were made in 1964 and 1965 by Voskhod spacecraft, which were modified Vostok spacecraft. By the late 1960s both were superseded by the Soyuz spacecraft, which are still used as of 2020.

The Vostok spacecraft was originally designed for use both as a camera platform (for the Soviet Union's first spy satellite program, Zenit) and as a crewed spacecraft. This dual-use design was crucial in gaining Communist Party support for the program. The basic Vostok design has remained in use for some 40 years, gradually adapted for a range of other uncrewed satellites. The descent module design was reused, in heavily modified form, by the Voskhod program.

Types of Spaceships: Past, Present, and Future

2.3 USA-Mercury

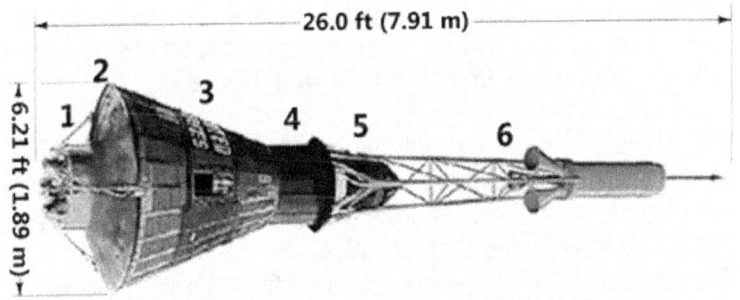

Project Mercury was the first human spaceflight program of the United States, running from 1958 through 1963. An early highlight of the Space Race, its goal was to put a man into Earth orbit and return him safely, ideally before the Soviet Union. Taken over from the US Air Force by the newly created civilian space agency NASA, it conducted twenty uncrewed developmental flights (some using animals), and six successful flights by astronauts. The program, which took its name from Roman mythology, cost $2.25 billion adjusted for inflation. The astronauts were collectively known as the "Mercury Seven", and each spacecraft was given a name ending with a "7" by its pilot.

The Space Race began with the 1957 launch of the Soviet satellite Sputnik 1. This came as a shock to the American public, and led to the creation of NASA to expedite existing US space exploration efforts, and place most of them under civilian control. After the successful launch of the Explorer 1 satellite in 1958, crewed spaceflight became the next goal. The Soviet Union put the first human, cosmonaut Yuri Gagarin, into a single orbit aboard Vostok 1 on April 12, 1961. Shortly after this, on May 5, the US launched its first astronaut, Alan Shepard, on a suborbital flight. Soviet Gherman Titov followed with a day-long

Types of Spaceships: Past, Present, and Future

orbital flight in August 1961. The US reached its orbital goal on February 20, 1962, when John Glenn made three orbits around the Earth. When Mercury ended in May 1963, both nations had sent six people into space, but the Soviets led the US in total time spent in space.

The Mercury space capsule was produced by McDonnell Aircraft, and carried supplies of water, food and oxygen for about one day in a pressurized cabin. Mercury flights were launched from Cape Canaveral Air Force Station in Florida, on launch vehicles modified from the Redstone and Atlas D missiles. The capsule was fitted with a launch escape rocket to carry it safely away from the launch vehicle in case of a failure. The flight was designed to be controlled from the ground via the Manned Space Flight Network, a system of tracking and communications stations; back-up controls were outfitted on board. Small retrorockets were used to bring the spacecraft out of its orbit, after which an ablative heat shield protected it from the heat of atmospheric reentry. Finally, a parachute slowed the craft for a water landing. Both astronaut and capsule were recovered by helicopters deployed from a US Navy ship.

The Mercury project gained popularity, and its missions were followed by millions on radio and TV around the world. Its success laid the groundwork for Project Gemini, which carried two astronauts in each capsule and perfected space docking maneuvers essential for crewed lunar landings in the subsequent Apollo program announced a few weeks after the first crewed Mercury flight.

2.4 USA-Gemini

Project Gemini was NASA's second human spaceflight program. Conducted between projects Mercury and Apollo, Gemini started in 1961 and concluded in 1966. The Gemini spacecraft carried a two-astronaut crew. Ten Gemini crews and sixteen individual astronauts flew low Earth orbit (LEO) missions during 1965 and 1966.

Gemini's objective was the development of space travel techniques to support the Apollo mission to land astronauts on the Moon. In doing so, it allowed the United States to catch up and overcome the lead in human spaceflight capability the Soviet Union had obtained in the early years of the Space Race, by demonstrating: mission endurance up to just under fourteen days, longer than the eight days required for a round trip to the Moon; methods of performing extra-vehicular activity (EVA) without tiring; and the orbital maneuvers necessary to achieve rendezvous and docking with another spacecraft. This left

Types of Spaceships: Past, Present, and Future

Apollo free to pursue its prime mission without spending time developing these techniques.

All Gemini flights were launched from Launch Complex 19 (LC-19) at Cape Kennedy Air Force Station in Florida. Their launch vehicle was the Gemini–Titan II, a modified Intercontinental Ballistic Missile (ICBM). Gemini was the first program to use the newly built Mission Control Center at the Houston Manned Spacecraft Center for flight control.

The astronaut corps that supported Project Gemini included the "Mercury Seven", "The New Nine", and the 1963 astronaut class. During the program, three astronauts died in air crashes during training, including both members of the prime crew for Gemini 9. This mission was flown by the backup crew.

Gemini was robust enough that the United States Air Force planned to use it for the Manned Orbital Laboratory (MOL) program, which was later canceled. Gemini's chief designer, Jim Chamberlin, also made detailed plans for cislunar and lunar landing missions in late 1961. He believed Gemini spacecraft could fly in lunar operations before Project Apollo, and cost less. NASA's administration did not approve those plans. In 1969, McDonnell-Douglas proposed a "Big Gemini" that could have been used to shuttle up to 12 astronauts to the planned space stations in the Apollo Applications Project (AAP). The only AAP project funded was Skylab – which used existing spacecraft and hardware – thereby eliminating the need for Big Gemini.

Types of Spaceships: Past, Present, and Future

2.5 Soviet-Voskhod

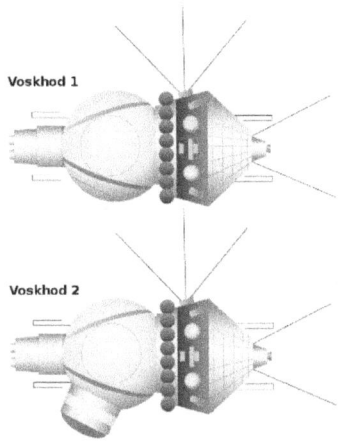

The Voskhod was a spacecraft built by the Soviet Union's space program for human spaceflight as part of the Voskhod program. It was a development of and a follow-on to the Vostok spacecraft. Voskhod 1 was used for a three-man flight whereas Voskhod 2 had a crew of two. They consisted of a spherical descent module (diameter 2.3 metres (7.5 ft)), which housed the cosmonauts, and instruments, and a conical equipment module (mass 2.27 tons or 5,000 pounds, 2.25 m (7.4 ft) long, 2.43 m (8.0 ft) wide), which contained propellant and the engine system. Voskhod was superseded by the Soyuz spacecraft in 1967.

Types of Spaceships: Past, Present, and Future

Types of Spaceships: Past, Present, and Future

2.6 USA-Apollo Command Module

The Apollo command and service module (CSM) was one of two principal components of the United States Apollo spacecraft, used for the Apollo program, which landed astronauts on the Moon between 1969 and 1972. The CSM functioned as a mother ship, which carried a crew of three astronauts and the second Apollo spacecraft, the Apollo Lunar Module, to lunar orbit, and brought the astronauts back to Earth. It consisted of two parts: the conical command module, a cabin that housed the crew and carried equipment needed for atmospheric reentry and splashdown; and the cylindrical service module which provided propulsion, electrical power and storage for various consumables required during a mission. An umbilical connection transferred power and consumables between the two modules. Just before reentry of the command module on the return home, the umbilical connection was severed and the service module was cast off and allowed to burn up in the atmosphere.

Types of Spaceships: Past, Present, and Future

The CSM was developed and built for NASA by North American Aviation starting in November 1961. It was initially designed to land on the Moon atop a landing rocket stage and return all three astronauts on a direct-ascent mission, which would not use a separate lunar module, and thus had no provisions for docking with another spacecraft. This, plus other required design changes, led to the decision to design two versions of the CSM: Block I was to be used for uncrewed missions and a single crewed Earth orbit flight (Apollo 1), while the more advanced Block II was designed for use with the lunar module. The Apollo 1 flight was cancelled after a cabin fire killed the crew and destroyed their command module during a launch rehearsal test. Corrections of the problems which caused the fire were applied to the Block II spacecraft, which was used for all crewed spaceflights.

Nineteen CSMs were launched into space. Of these, nine flew humans to the Moon between 1968 and 1972, and another two performed crewed test flights in low Earth orbit, all as part of the Apollo program. Before these, another four CSMs had flown as uncrewed Apollo tests, of which two were suborbital flights and another two were orbital flights. Following the conclusion of the Apollo program and during 1973–1974, three CSMs ferried astronauts to the orbital Skylab space station. Finally in 1975, the last flown CSM docked with the Soviet craft Soyuz 19 as part of the international Apollo–Soyuz Test Project.

Types of Spaceships: Past, Present, and Future

2.7 USA-Space Shuttle Orbiter

The Space Shuttle orbiter is the spaceplane component of the Space Shuttle, a partially reusable orbital spacecraft system that was part of the Space Shuttle program. Operated by NASA, the U.S. space agency, this vehicle could carry astronauts and payloads into low Earth orbit, perform in-space operations, then re-enter the atmosphere and land as a glider, returning its crew and any on-board payload to the Earth.

Six orbiters were built for flight: Enterprise, Columbia, Challenger, Discovery, Atlantis, and Endeavour. All were built in Palmdale, California, by the Pittsburgh, Pennsylvania-based Rockwell International company. The first orbiter, Enterprise, made its maiden flight in 1977. An unpowered glider, it was carried by a modified Boeing 747 airliner called the Shuttle Carrier Aircraft and released for a series of atmospheric test flights and landings. Enterprise was partially disassembled and retired after completion of critical testing. The remaining orbiters were fully operational spacecraft, and were launched vertically as part of the Space Shuttle stack.

Columbia was the first space-worthy orbiter, and made its inaugural flight in 1981. Challenger, Discovery, and Atlantis followed in 1983, 1984 and 1985 respectively. In 1986, Challenger was destroyed in an accident shortly after launch. Endeavour was built as Challenger's replacement, and was first launched in 1992. In 2003, Columbia was destroyed during re-entry, leaving just three remaining orbiters. Discovery completed its final flight on March 9, 2011, and Endeavour completed its final flight on June 1, 2011. Atlantis completed the last ever Shuttle flight, STS-135, on July 21, 2011.

In addition to their crews and payloads, the reusable orbiter carried most of the Space Shuttle System's liquid-fueled rocket propulsion system, but both the liquid hydrogen fuel and the liquid oxygen oxidizer for its three main rocket engines were fed from an external cryogenic propellant tank. Additionally, two reusable solid rocket boosters provided additional thrust for approximately the first two minutes of launch. The orbiters themselves did carry hypergolic propellants for their RCS thrusters and Orbital Maneuvering System engines.

Types of Spaceships: Past, Present, and Future

2.8 Soviet-Buran Shuttle

Buran meaning "Snowstorm" or "Blizzard"; GRAU index serial number: 11F35 1K, construction number: 1.01) was the first spaceplane to be produced as part of the Soviet/Russian Buran program. Besides describing the first operational Soviet/Russian shuttle orbiter, "Buran" was also the designation for the entire Soviet/Russian spaceplane project and its orbiters, which were known as "Buran-class orbiters".

Buran completed one uncrewed spaceflight in 1988, and was destroyed in 2002 when the hangar it was stored in collapsed. The Buran-class orbiters used the expendable Energia rocket, a class of super heavy-lift launch vehicle.

Types of Spaceships: Past, Present, and Future

The only orbital launch of a Buran-class orbiter, 1K1 (first orbiter, first flight) occurred at 03:00:02 UTC on 15 November 1988 from Baikonur Cosmodrome launch pad 110/37. Buran was lifted into space, on an uncrewed mission, by the specially designed Energia rocket. The automated launch sequence performed as specified, and the Energia rocket lifted the vehicle into a temporary orbit before the orbiter separated as programmed. After boosting itself to a higher orbit and completing two orbits around the Earth, the ODU engines fired automatically to begin the descent into the atmosphere, return to the launch site, and horizontal landing on a runway.

After making an automated approach to Site 251, Buran touched down under its own control at 06:24:42 UTC and came to a stop at 06:25:24, 206 minutes after launch. Despite a lateral wind speed of 61.2 kilometers' per hour (38.0 mph), Buran landed only 3 metres (9.8 ft) laterally and 10 metres (33 ft) longitudinally from the target mark. It was the first spaceplane to perform an uncrewed flight, including landing in fully automatic mode. It was later found that Buran had lost only eight of its 38,000 thermal tiles over the course of its flight.

Types of Spaceships: Past, Present, and Future

3.0 Current Spaceships

Today there are some workhorse spaceships like the Soyuz which have flown for decades and some new contenders like the Space X Falcon 9 rockets which recently launched two astronauts to the International Space Station

3.1 Soviet/Russia-Soyuz

Soyuz is a series of spacecraft designed for the Soviet space program by the Korolev Design Bureau (now RKK Energia) in the 1960s that remains in service today, having made more than 140 flights. The Soyuz succeeded the Voskhod spacecraft and was originally built as part of the Soviet crewed lunar programs. The Soyuz spacecraft is launched on a Soyuz rocket, the most reliable launch vehicle in the world to date. The Soyuz rocket design is based on the Vostok launcher, which in turn was based on the 8K74 or R-7A Semyorka, a Soviet intercontinental ballistic missile. All Soyuz spacecraft are launched from the Baikonur Cosmodrome in Kazakhstan. After the retirement of the Space Shuttle in 2011, the Soyuz served as the only means for Americans to make crewed space flights until the first flight of VSS Unity in 2018, and the only means for Americans to reach the International Space Station until the first flight of Dragon 2 Crew variant on May 30, 2020. The Soyuz is heavily used in the ISS program.

The first Soyuz flight was uncrewed and started on November 28, 1966. The first Soyuz mission with a crew, Soyuz 1, launched on 23 April 1967 but ended with a crash due to a parachute failure, killing cosmonaut Vladimir Komarov. The following flight was uncrewed. Soyuz 3, launched on October 26, 1968, became the program's first successful crewed mission.

The only other flight to suffer a fatal accident, Soyuz 11, killed its crew of three when the cabin depressurized prematurely just before reentry. These were the only humans to date to have died above the Kármán line. Despite these early incidents, Soyuz is widely considered the world's safest, most cost-effective human spaceflight vehicle established by its unparalleled length of operational history. Soyuz spacecraft were used to carry cosmonauts to and from Salyut and later Mir Soviet space stations, and are now used for transport to and from the International Space Station (ISS). At least one Soyuz spacecraft is docked to ISS at all times for use as an escape craft in the event of an emergency. The spacecraft is intended to be replaced by the six-person Orel spacecraft.

3.2 USA-Space X Dragon Manned

SPACEX CREW DRAGON
DM-2 Launch Configuration

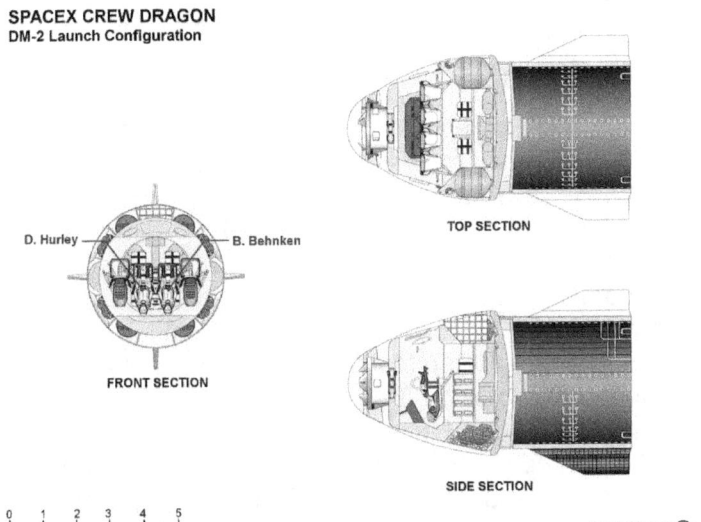

The SpaceX Dragon 2 is a class of reusable spacecraft developed and manufactured by American aerospace manufacturer SpaceX as the successor to Dragon, a reusable cargo spacecraft. It has two variants: Crew Dragon, a space capsule capable of ferrying up to seven astronauts, and Cargo Dragon, an updated replacement for the original Dragon spacecraft. The spacecraft launches atop a Falcon 9 Block 5 rocket and returns to Earth via an ocean splashdown. Unlike its predecessor, the spacecraft can dock itself to the ISS instead of being berthed. Crew Dragon is equipped with an integrated launch escape system (LES) capable of accelerating the vehicle away from the rocket in an emergency at 11.8 m/s2 (39 ft/s2), accomplished by using a set of four side-mounted thruster pods with two SuperDraco engines each. The spacecraft features redesigned solar arrays and a modified outer mold line compared to the original Dragon, and possesses new flight computers and avionics. As of March 2020, four Dragon 2 spacecraft have been manufactured (not counting structural test articles that were never airborne).

Crew Dragon serves as one of two spacecraft that is expected to transport crews to and from the International Space Station (ISS) under NASA's Commercial Crew Program, the other being the Boeing CST-100 Starliner. It is also expected to be used in flights by American space tourism company Space Adventures and to shuttle tourists to and from Axiom Space's planned space station. Crew Dragon's first non-piloted test flight occurred in March 2019, and its first crewed flight – with astronauts Robert Behnken and Douglas Hurley – occurred in May 2020. This test flight marked the first time a private company launched a crewed orbital spacecraft. Cargo Dragon is expected to supply cargo to the ISS under a Commercial Resupply Services-2 contract with NASA, along with Northrop Grumman Innovation Systems' Cygnus spacecraft and Sierra Nevada Corporation's Dream Chaser spacecraft. The first flight of the Cargo Dragon is planned to launch in October 2020.

There are two variants: Crew Dragon and Cargo Dragon. Crew Dragon was initially called "DragonRider" and it was intended from the beginning to support a crew of seven or a combination of crew and cargo. It is able to perform fully autonomous rendezvous and docking with manual override ability, using the NASA Docking System (NDS). For typical missions, Crew Dragon will remain docked to the ISS for a period of 180 days, but is designed to remain on the station for up to 210 days, matching the Russian Soyuz spacecraft. From the beginning of the development process, SpaceX planned to use an integrated pusher launch escape system for the Dragon spacecraft.

Types of Spaceships: Past, Present, and Future

The SpaceX Dragon 2 Control Panel

Types of Spaceships: Past, Present, and Future

3.3 China-Shenzou

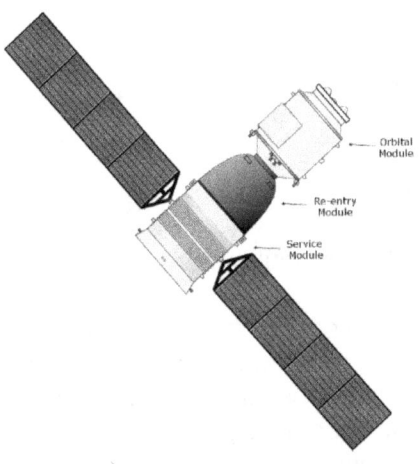

Shenzhou is a spacecraft developed and operated by China using Soyuz technology to support its crewed spaceflight program. The name is variously translated as Divine vessel, Divine craft, or Divine ship. Its design resembles the Russian Soyuz spacecraft, but it is larger in size. The first launch was on 19 November 1999 and the first crewed launch was on 15 October 2003. In March 2005, an asteroid was named 8256 Shenzhou in honor of the spacecraft.

Shenzhou consists of three modules: a forward orbital module, a reentry module in the middle, and an aft service module. This division is based on the principle of minimizing the amount of material to be returned to Earth. Anything placed in the orbital or service modules does not require heat shielding, increasing the space available in the spacecraft without increasing weight as much as it would if those modules were also able to withstand reentry.

Types of Spaceships: Past, Present, and Future

Complete spacecraft data

Total mass: 7840 kg
Length: 9.25 m
Diameter: 2.80 m
Span: 17.00 m

Orbital module

Shenzhou's Orbital Module prior to S8

The orbital module contains space for experiments, crew-serviced or crew-operated equipment, and in-orbit habitation. Without docking systems, Shenzhou 1–6 carried different kinds of payload on the top of their orbital modules for scientific experiments.

Up until Shenzhou 8, the orbital module of the Shenzhou was equipped with its own propulsion, solar power, and control systems, allowing autonomous flight. It is possible for Shenzhou to leave an orbital module in orbit for redocking with a later spacecraft, a capability which Soyuz does not possess, since the only hatch between the orbital and reentry modules is a part of the reentry module, and orbital module is depressurized after separation. For future missions, the orbital module(s) could also be left behind on the planned Chinese project 921/2 space station as additional station modules.

In the uncrewed test flights launched, the orbital module of each Shenzhou was left functioning on orbit for several days after the reentry modules return, and the Shenzhou 5 orbital module continued to operate for six months after launch.

Types of Spaceships: Past, Present, and Future

Orbital module data

Design life: 200 days
Length: 2.80 m (9.10 ft)
Basic diameter: 2.25 m (7.38 ft)
Maximum diameter: 2.25 m (7.38 ft)
Span: 10.40 m (34.10 ft)
Habitable volume: 8.00 m³
Mass: 1500 kg (3,300 lb)
RCS Coarse No x Thrust: 16 x 5 N
RCS Propellants: Hydrazine
Electrical system: Solar panels, 12.24 m²
Electric system: 0.50 average kW
Electric system: 1.20 kW

Reentry module

The reentry module is located in the middle section of the spacecraft and contains seating for the crew. It is the only portion of Shenzhou which returns to Earth's surface. Its shape is a compromise between maximizing living space and allowing for some aerodynamic control upon reentry.

Reentry module data

Crew size: 3
Design life: 20 days
Length: 2.50 m (8.20 ft)
Basic diameter: 2.52 m (8.26 ft)
Maximum diameter: 2.52 m (8.26 ft)
Habitable volume: 6.00 m³
Mass: 3240 kg (7,140 lb)
Heat shield mass: 450 kg (990 lb)
Lift-to-drag-ratio (hypersonic): 0.30
RCS Coarse No x Thrust: 8 x 150 N
RCS Propellants: Hydrazine

Types of Spaceships: Past, Present, and Future

Service module

The aft service module contains life support and other equipment required for the functioning of Shenzhou. Two pairs of solar panels, one pair on the service module and the other pair on the orbital module, have a total area of over 40 m2 (430 ft²), indicating average electrical power over 1.5 kW (Soyuz have 1.0 kW).

Service module data

Design life: 20 days
Length: 2.94 m (9.65 ft)
Basic diameter: 2.50 m (8.20 ft)
Maximum diameter: 2.80 m (9.10 ft)
Span: 17.00 m (55.00 ft)
Mass: 3000 kg (6,600 lb)
RCS Coarse No x Thrust: 8 x 150 N
RCS Fine No x Thrust: 16 x 5 N
RCS Propellants: N2O4 / MMH, unified system with main engine
Main engine: 4 x 2500 N
Main engine thrust: 10.000 kN (2,248 lbf)
Main engine propellants: N2O4 / MMH
Main engine propellants: 1000 kg (2,200 lb)
Main engine Isp: 290 seconds
Electrical system: Solar panels, 24.48 + 12.24 m², 36.72 m² total
Electric system: 1.50 average kW
Electric system: 2.40 kW

Types of Spaceships: Past, Present, and Future

4.0 Spacecraft in Progress

Quite a few spacecraft are still in the design, construction, and testing stages. These will be part of our space transportation systems in the near future.

4.1 USA-Boeing Starliner

Boeing Starliner (officially CST-100 Starliner) is a class of reusable crew capsules expected to transport crew to the International Space Station (ISS) and to private space stations such as the proposed Bigelow Aerospace Commercial Space Station. It is manufactured by Boeing for its participation in NASA's Commercial Crew Program.

The capsule has a diameter of 4.56 m (15.0 ft), which is slightly larger than the Apollo command module and smaller than the Orion capsule. The Boeing Starliner holds a crew of up to seven people and is being designed to be able to remain in-orbit for up to seven months with reusability of up to ten missions. It is designed to be compatible with four launch vehicles: Atlas V, Delta IV, Falcon 9, and Vulcan.

In the first phase of its CCP, NASA awarded Boeing US$18 million in 2010 for preliminary development of the spacecraft. In the second phase Boeing was awarded a US$93 million contract in 2011 for further spacecraft development. On 3 August 2012, NASA announced the award of US$460 million to Boeing to continue work on the Starliner under the Commercial Crew Integrated Capability (CCiCap) Program. On 16 September 2014, NASA selected the Boeing Starliner, along with SpaceX Crew Dragon, for the Commercial Crew Transportation Capability (CCtCap) program, with an award of US$4.2 billion.

The Boeing Starliner Orbital Flight Test (uncrewed test flight) launched with the Atlas V N22, on 20 December 2019 from SLC-41 at Cape Canaveral, Florida. During the test, the Starliner experienced a timing anomaly that precluded a docking with the International Space Station. Two days after launch, on 22 December 2019 at 12:58 UTC, with the successful landing at White Sands Missile Range, New Mexico, the Boeing Starliner Calypso became the first-ever crew-capable space capsule to make a land-based touchdown in the United States.

The design draws upon Boeing's experience with NASA's Apollo, Space Shuttle and ISS programs as well as the Orbital Express project sponsored by the Department of Defense. Starliner has no Orion heritage, but it is sometimes confused with the earlier and similar Orion-derived Orion Lite proposal that Bigelow Aerospace was reportedly working on with technical assistance from

Types of Spaceships: Past, Present, and Future

Lockheed Martin. It will use the NASA Docking System for docking and use the Boeing Lightweight Ablator for its heat shield. The Starliner's solar cells will provide more than 2.9 kW of electricity, and will be placed on top of the micro-meteoroid debris shield located at the bottom of the spacecraft's service module.

It is designed to be compatible with multiple launch vehicles, including the Atlas V, Delta IV, and Falcon 9, as well as the planned Vulcan.

Unlike earlier U.S. space capsules, Starliner will make airbag-cushioned landings on the ground rather than into water. Five landing areas are planned in the Western United States, which will give the Starliner about 450 landing opportunities every year.

Starliner includes one space tourist seat, and the Boeing contract with NASA allows Boeing to price and sell passage to low Earth orbit on that seat.

Types of Spaceships: Past, Present, and Future

4.2 USA-NASA Orion

Orion (officially Orion Multi-Purpose Crew Vehicle or Orion MPCV) is a class of partially reusable space capsules to be used in NASA's human spaceflight programs. The spacecraft consists of a Crew Module (CM) manufactured by Lockheed Martin and the European Service Module (ESM) manufactured by Airbus Defense and Space.

Capable of supporting a crew of six beyond low Earth orbit, Orion can last up to 21 days undocked and up to six months docked. It is equipped with solar panels, an automated docking system, and glass cockpit interfaces modeled after those used in the Boeing 787 Dreamliner. A single AJ10 engine provides the spacecraft's primary propulsion, while eight R-4D-11 engines, and six pods of custom reaction control system engines developed by Airbus, provide the spacecraft's secondary propulsion. Although compatible with other launch vehicles, Orion is primarily designed to launch atop a Space Launch System (SLS) rocket, with a tower launch escape system.

Types of Spaceships: Past, Present, and Future

Orion was originally conceived by Lockheed Martin as a proposal for the Crew Exploration Vehicle (CEV) to be used in NASA's Constellation program. Lockheed Martin's proposal defeated a competing proposal by Northrop Grumman, and was selected by NASA in 2006 to be the CEV. Originally designed with a service module featuring a new "Orion Main Engine" and a pair of circular solar panels, the spacecraft was to be launched atop the Ares I rocket. Following the cancellation of the Constellation program in 2010, Orion was heavily redesigned for use in NASA's Journey to Mars initiative; later named Moon to Mars. The SLS replaced the Ares I as Orion's primary launch vehicle, and the service module was replaced with a design based on the European Space Agency's Automated Transfer Vehicle. A development version of Orion's CM was launched in 2014 during Exploration Flight Test-1, while at least four test articles have been produced. As of 2020, three flight-worthy Orion spacecraft are under construction, with an additional one ordered, for use in NASA's Artemis program; the first of these is due to be launched in 2021 on Artemis 1.

The Orion crew module (CM) is a reusable transportation capsule that provides a habitat for the crew, provides storage for consumables and research instruments, and contains the docking port for crew transfers. The crew module is the only part of the spacecraft that returns to Earth after each mission and is a 57.5° truncated cone shape with a blunt spherical aft end, 5.02 meters (16 ft 6 in) in diameter and 3.3 meters (10 ft 10 in) in length, with a mass of about 8.5 metric tons (19,000 lb). It was manufactured by the Lockheed Martin Corporation at Michoud Assembly Facility in New Orleans. It will have 50% more volume than the Apollo capsule and will carry four to six astronauts. After extensive study, NASA has

Types of Spaceships: Past, Present, and Future

selected the Avcoat ablator system for the Orion crew module. Avcoat, which is composed of silica fibers with a resin in a honeycomb made of fiberglass and phenolic resin, was formerly used on the Apollo missions and on the Space Shuttle orbiter for early flights.

Orion's CM will use advanced technologies, including:

Glass cockpit digital control systems derived from those of the Boeing 787.

An "autodock" feature, like those of Progress, the Automated Transfer Vehicle, and Dragon 2, with provision for the flight crew to take over in an emergency. Prior US spacecraft have all been docked by the crew.

Improved waste-management facilities, with a miniature camping-style toilet and the unisex "relief tube" used on the Space Shuttle.

A nitrogen/oxygen (N2/O2) mixed atmosphere at either sea level (101.3 kPa or 14.69 psi) or reduced (55.2 to 70.3 kPa or 8.01 to 10.20 psi) pressure.

Far more advanced computers than on prior crew vehicles. The CM will be built of aluminium-lithium alloy. The reusable recovery parachutes will be based on the parachutes used on both the Apollo spacecraft and the Space Shuttle Solid Rocket Boosters, and will be constructed of Nomex cloth. Water landings will be the exclusive means of recovery for the Orion CM.

To allow Orion to mate with other vehicles, it will be equipped with the NASA Docking System. The spacecraft will employ a Launch Escape System (LES) along with a "Boost Protective Cover" (made of fiberglass), to protect the Orion CM from aerodynamic and impact stresses during the first 2 1⁄2 minutes of ascent. Its designers claim that the MPCV is designed to be 10 times safer during ascent and reentry than the Space Shuttle. The CM is designed to be refurbished and reused. In addition, all of Orion's component parts have been designed to be as modular as possible, so that between the craft's first test flight in 2014 and its projected Mars voyage in the 2030s, the spacecraft can be upgraded as new technologies become available.

As of 2019, the Spacecraft Atmospheric Monitor is planned to be used in the Orion CM

Orion Control Panels

Types of Spaceships: Past, Present, and Future

4.3 USA-Virgin Galactic-Spaceship Two

The Scaled Composites Model 339 SpaceShipTwo (SS2) is an air-launched suborbital spaceplane type designed for space tourism. It is manufactured by The Spaceship Company, a California-based company owned by Virgin Galactic.

SpaceShipTwo is carried to its launch altitude by a Scaled Composites White Knight Two, before being released to fly on into the upper atmosphere powered by its rocket engine. It then glides back to Earth and performs a conventional runway landing. The spaceship was officially unveiled to the public on 7 December 2009 at the Mojave Air and Space Port in California. On 29 April 2013, after nearly three years of unpowered testing, the first one constructed successfully performed its first powered test flight.

Virgin Galactic plans to operate a fleet of five SpaceShipTwo spaceplanes in a private passenger-carrying service and has been taking bookings for some time, with a suborbital flight carrying an updated ticket price of US $250,000. The spaceplane could also be used to carry scientific payloads for NASA and other organizations.

On 31 October 2014, during a test flight, the first SpaceShipTwo VSS Enterprise broke up in flight and crashed in the Mojave Desert. A preliminary investigation suggested that the craft's descent device deployed too early. One pilot, Michael Alsbury, was killed; the other was treated for a serious shoulder injury after parachuting from the stricken spacecraft.

The second SpaceShipTwo spacecraft, VSS Unity, was unveiled on 19 February 2016. The vehicle is undergoing flight testing. Its first flight to space (above 50 miles altitude), VSS Unity VP03, took place on 13 December 2018.

The SpaceShipTwo project is based in part on technology developed for the first-generation SpaceShipOne, which was part of the Scaled Composites Tier One program, funded by Paul Allen. The Spaceship Company licenses this technology from Mojave Aerospace Ventures, a joint venture of Paul Allen and Burt Rutan, the designer of the predecessor technology.

SpaceShipTwo is a low-aspect-ratio passenger spaceplane. Its capacity will be eight people — six passengers and two pilots. The apogee of the new craft will be approximately 110 km (68 mi) in the lower thermosphere, 10 km (6.2 mi) higher than the Kármán line which was SpaceShipOne's target, although the last flight of SpaceShipOne reached a one-time altitude of 112 km (70 mi). SpaceShipTwo will reach 4,200 km/h (2,600 mph),

Types of Spaceships: Past, Present, and Future

using a single hybrid rocket engine — the RocketMotorTwo. It launches from its mother ship, White Knight Two, at an altitude of 15,000 metres (50,000 ft), and reaches supersonic speed within 8 seconds. After 70 seconds, the rocket engine cuts out and the spacecraft will coast to its peak altitude. SpaceShipTwo's crew cabin is 3.7 m (12 ft) long and 2.3 m (7.5 ft) in diameter. The wing span is 8.2 m (27 ft), the length is 18 m (60 ft) and the tail height is 4.6 m (15 ft).

SpaceShipTwo uses a feathered reentry system, feasible due to the low speed of reentry. In contrast, orbital spacecraft re-enter at orbital speeds, close to 25,000 km/h (16,000 mph), using heat shields. SpaceShipTwo is furthermore designed to re-enter the atmosphere at any angle. It will decelerate through the atmosphere, switching to a gliding position at an altitude of 24 km (15 mi), and will take 25 minutes to glide back to the spaceport.

SpaceShipTwo and White Knight Two are, respectively, roughly twice the size of the first-generation SpaceShipOne and mothership White Knight, which won the Ansari X Prize in 2004. SpaceShipTwo has 43 and 33 cm (17 and 13 in) -diameter windows for the passengers' viewing pleasure, and all seats will recline back during landing to decrease the discomfort of G-forces. Reportedly, the craft can land safely even if a catastrophic failure occurs during flight. In 2008, Burt Rutan remarked on the safety of the vehicle:

This vehicle is designed to go into the atmosphere in the worst case straight in or upside down and it'll correct. This is designed to be at least as safe as the early airliners in the 1920s ... Don't believe anyone that tells you that the safety will be the same as a modern airliner, which has been around for 70 years.

Types of Spaceships: Past, Present, and Future

In September 2011, the safety of SpaceShipTwo's feathered reentry system was tested when the crew briefly lost control of the craft during a gliding test flight. Control was reestablished after the spaceplane entered its feathered configuration, and it landed safely after a 7-minute flight.

4.4 USA-Blue Origin-New Shepard

The New Shepard is a fully reusable, vertical takeoff, vertical landing (VTVL) space vehicle composed of two principal parts: a pressurized crew capsule and a booster rocket that Blue Origin calls a propulsion module. The New Shepard is controlled entirely by on-board computers, without ground control or a human pilot.

Crew capsule

The New Shepard Crew Capsule is a pressurized crew capsule that can carry six people, and supports a "full-envelope" launch escape system that can separate the capsule from the booster rocket anywhere during the ascent. Interior volume of the capsule is 15 cubic meters (530 cu ft). The Crew Capsule Escape Solid Rocket Motor (CCE-SRM) is sourced from Aerojet Rocketdyne. After separation two or three parachutes deploy. Just before landing, retro rockets fire.

Types of Spaceships: Past, Present, and Future

Propulsion module

The New Shepard propulsion module is powered using a Blue Origin BE-3 bipropellant rocket engine burning liquid hydrogen and liquid oxygen, although some early development work was done by Blue Origin on engines operating with other propellants: the BE-1 engine using monopropellant hydrogen peroxide; and the BE-2 engine using high-test peroxide oxidizer and RP-1 kerosene fuel.

Types of Spaceships: Past, Present, and Future

4.5 USA-Sierra Nevada-Dream Chaser

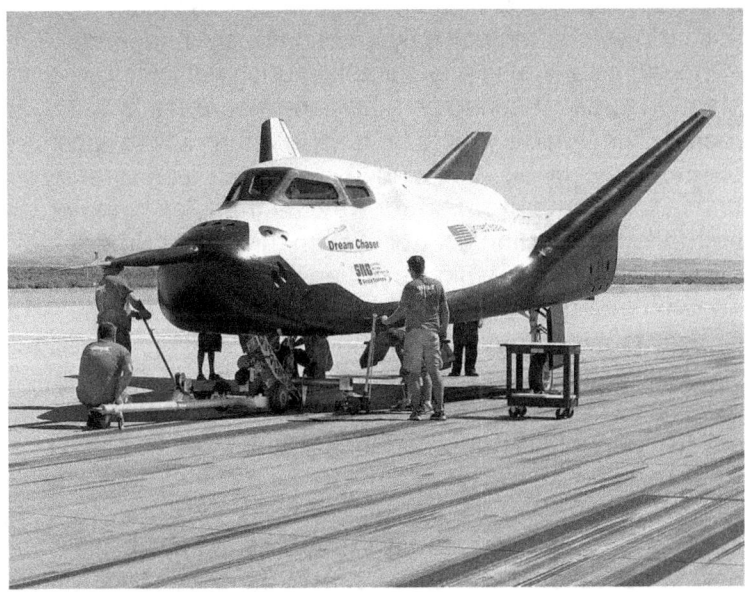

The Dream Chaser Cargo System is an American reusable lifting body spaceplane being developed by Sierra Nevada Corporation (SNC) Space Systems. Originally intended as a crewed vehicle, the Dream Chaser Space System, to be produced after the cargo variant is operational, is capable of carrying up to seven people and cargo to and from low Earth orbit.

The cargo Dream Chaser will resupply the International Space Station with both pressurized and unpressurized cargo. It will be launched vertically on the Vulcan Centaur rocket, and autonomously land horizontally on conventional runways. A proposed version operated by ESA would launch on Ariane 5.

The originally planned Dream Chaser Space System is a human-rated version designed to carry from two to seven people and cargo to orbital destinations such as the International Space Station. It was to have a built-in launch escape system and could fly autonomously if needed. Although it could use any suitable launch vehicle, it was planned to be launched on a human-rated Atlas V 412 rocket. The vehicle was to be able to return from space by gliding (typically experiencing less than 1.5 g on re-entry) and landing on any airport runway that handles commercial air traffic. Its reaction control system thrusters burned ethanol-based fuel, which is not an explosively volatile material, nor toxic like hydrazine, allowing the Dream Chaser to be handled immediately after landing, unlike the Space Shuttle. Its thermal protection system (TPS) was made up of silica-based tiles and a new composite material called Toughened Unipiece Fibrous Reusable Oxidation Resistant Ceramic (TUFROC).

As of 2020, the Sierra Nevada Corporation says it still plans to produce a crewed version of the spacecraft within the next 5 years. The company says it "never stopped working" on the crewed version and fully intends to launch it after the cargo version.

Types of Spaceships: Past, Present, and Future

5.0 Future Spaceships in Design

There are also some very active projects to build the next generation of spaceships like the Space X Starship which will be fully reusable and will hopefully launch humans to Mars someday.

5.1 USA-SpaceX-Starship

The SpaceX Starship system is a fully-reusable, two-stage-to-orbit, super heavy-lift launch vehicle under development by SpaceX since 2012, as a self-funded private spaceflight project.

The second stage—which is also referred to as "Starship":16:20–16:48—is being designed as a long-duration cargo, and eventually,, passenger-carrying spacecraft. It is being used initially without any booster stage at all, as part of an extensive development program to prove out launch-and-landing and iterate on a variety of design details, particularly with respect to the vehicle's atmospheric reentry. While the spacecraft is currently being tested on its own at suborbital altitudes during 2019–20, it will later be used on orbital launches with an additional booster stage, the Super Heavy, where the spacecraft will serve as both the second stage on the two-stage-to-orbit launch vehicle and the in-space long-duration orbital spaceship.

Integrated system testing of a proof of concept for Starship began in March 2019, with the addition of a single Raptor rocket engine to a reduced-height prototype, nicknamed Starhopper — similar to Grasshopper, an equivalent prototype of the Falcon 9 reusable booster. Starhopper was used from April through August 2019 for static testing and low-altitude, low-velocity flight testing of vertical launches and landings in July and August 2019. More prototype Starships have been built and more are under construction as the iterative design goes through several iterations. All test articles have a 9-meter (30 ft)-diameter stainless steel hull.

SpaceX is planning to launch commercial payloads using Starship no earlier than 2021.In April 2020, NASA selected a modified human-rated Starship system as one of three potential lunar landing system design concepts to receive funding for a 10-month long initial design phase for the NASA Artemis program.

The launch vehicle was initially mentioned in public discussions by SpaceX CEO Elon Musk in 2012 as part of a description of the company's overall Mars system architecture, then known as "Mars Colonial Transporter" (MCT). By August 2014, media sources speculated that the initial flight test of the Raptor-driven super-heavy launch vehicle could occur as early as 2020, in order to fully test the engines under orbital spaceflight conditions; however, any colonization effort was then reported to continue to be "deep into the future".

In mid-September 2016, Musk noted that the Mars Colonial Transporter name would not continue, as the system would be able to "go well beyond Mars", and that a new name would be needed. The name selected was "Interplanetary Transport System" (ITS), although in an AMA on Reddit on October 23, 2016, Musk stated, "I think we need a new name. ITS just isn't working. I'm using BFR and BFS for the rocket and spaceship, which is fine internally, but...", without stating what the new name might be. In September 2017, at the 68th annual meeting of the International Astronautical Congress, SpaceX unveiled an updated vehicle design. Musk said, "we are searching for the right name, but the code name, at least, is BFR".

In a September 2018 announcement of a planned 2023 lunar circumnavigation mission, a private spaceflight called dearMoon project, Musk showed another redesigned concept for the second stage and spaceship with three rear fins and two front canard fins added for atmospheric entry, replacing the previous delta wing and split flaps shown a year earlier. The design was to use seven identically-sized Raptor engines in the second stage; the same engine model as would be used on the first stage. The second stage design had two small actuating canard fins near the nose of the ship, and three large fins at the base, two of which would actuate, with all three serving as landing legs. Additionally, SpaceX also stated in the second half of the month that they were "no longer planning to upgrade Falcon 9 second stage for reusability". The two major parts of the launch vehicle were given descriptive names in November 2018: "Starship" for the upper stage and "Super Heavy" for the booster stage, which Musk pointed out was "needed to escape Earth's deep gravity well (not needed for other planets or moons)".

In January 2019, Elon Musk announced that the Starship would no longer be constructed out of carbon fiber, and that stainless steel would be used instead to build the Starship. Musk cited several reasons including cost, strength, and ease of production to justify making the switch.

In May 2019, the Starship design changed back to just six Raptor engines, with three optimized for sea-level and three optimized for vacuum. By late May 2019, an initial prototype, Starhopper, was being finished for untethered flight tests at the SpaceX South Texas launch site, while two "orbital prototypes" were under construction, one in South Texas begun in March 2019 and one on the Florida space coast begun before May 2019. The build of the first Super Heavy booster stage was projected to be able to start by September 2019. At the time, neither of the two orbital prototypes yet had aerodynamic control surfaces nor landing legs added to the under construction tank structures, and Musk indicated that the design for both would be changing once again. In September 2019, the externally-visible "moving fins" began to be added to the Mk1 prototype, giving a view into the promised mid-2019 redesign of the aerodynamic control surfaces for the test vehicles.

In June 2019, SpaceX publicly announced discussions had begun with three telecommunications companies for using Starship, rather than Falcon 9, for launching commercial satellites for paying customers in 2021. No specific companies or launch contracts were announced at that time.

Types of Spaceships: Past, Present, and Future

In July 2019, the Starhopper made its initial flight test, a "hop" of around 20 m (66 ft) altitude, and a second and final "hop" in August 2019, reached an altitude of ~150 m (490 ft) and landing around 100 m (330 ft) from the Launchpad.

SpaceX completed most of the Boca Chica prototype, the Starship Mk1, in time for Musk's next public update in September 2019. Watching the construction in progress before the event, observers online circulated photos and speculated about the most visible change, a move to two tail fins from the earlier three. During the event, Musk added that landing would now be accomplished on six dedicated landing legs, following a re-entry protected by glass heat tiles. Updated specifications were provided: when optimized, Starship was expected to mass at 120,000 kg (260,000 lb) empty and be able to initially transport a payload of 100,000 kg (220,000 lb) with an objective of growing that to 150,000 kg (330,000 lb) over time. Musk suggested that an orbital flight might be achieved by the fourth or fifth test prototype in 2020, using a Super Heavy booster in a two-stage-to-orbit launch vehicle configuration, and emphasis was placed on possible future lunar missions.

In September 2019, Elon Musk unveiled Starship Mk1. At the time, construction of Mk3 was expected to start in about a month.

In 2019, the cost per launch for Starship was estimated by SpaceX to be as low as US$2 million once the company achieves a robust operational cadence and achieves the technological advance of full and rapid reusability. Full reusability of the second stage of Starship is a fundamental design goal for the entire Starship development program, but success is uncertain. Elon Musk has said in 2020 that, with a high flight rate, they could potentially go even lower, with a fully-burdened marginal cost on the order of US$10 per kilogram of payload launched to low Earth orbit.

In November 2019, the Mk1 test article in Texas came apart in a tank pressure test, and SpaceX stated they would cease to build the Mk2 prototype under construction in Florida and move on to work on the Mk3 article. A few weeks later, the work on the vehicles in Florida slowed down substantially, with some assemblies that had been built in Florida for those vehicles being transported to the Texas Starship assembly location, and a reported 80% reduction in the workforce at the Florida assembly location as SpaceX pauses activities there. Apparently, at the same time the Mk4 vehicle under construction in Florida was cancelled.

The Mk3 article was renamed Starship SN1 by SpaceX, to signify the major evolution in building techniques: the rings were now taller and each was made of one single sheet of steel, drastically reducing the welding lines (thus failure points). Also the worksite was expanded with more tents, structures and upgraded machinery to give the workers a much better controlled environment, allowing for better precision work and, more importantly, better quality welds. Smaller articles were made and tested to destruction to validate the new bulkheads design. This was a significant change of pace for SpaceX's approach.

In February 2020, SN1 too was destroyed when undergoing pressurization. The company then focused on resolving the problem that led to SN1's failure by assembling a stripped down version of their next planned prototype, SN2. This time the test was successful and

Types of Spaceships: Past, Present, and Future

SpaceX began work on SN3. However, in April 2020, SN3 was also destroyed during testing due to a test configuration error. At that time, construction of SN4 was underway.

On 26 April 2020, Starship SN4 became the first full-scale prototype to pass a cryogenic proof test, in which the ship's liquid oxygen and methane was replaced with similarly frigid but non-explosive liquid nitrogen. SN4 was only pressurized to 4.9 bar (~70 psi), which is more than enough to perform a small flight test. On 5 May 2020, SN4 completed a single engine static fire with one mounted Raptor engine and became the first full Starship tank to pass a Raptor static fire. SN4 would complete a total of 4 short static fires (2 to 5 seconds long) before being destroyed in a massive explosion due to a propellant leak from the quick disconnect mechanism occurring at the end of a test fire on 29 May 2020.

SN7 test tank was also tested all the way to leak at around 7.8 bar after which it was repaired and tested to destruction.

On 30 July 2020, Starship SN5 completed a static fire with one mounted Raptor engine. On 3 August 2020, Starship SN5 attempted a 150 meter flight test that was aborted due to a spin start valve on the Raptor engine not opening. At approximately 11:00 AM CDT on 4 August 2020, SpaceX again attempted a 150 meter flight test, which was also aborted. Later that day, at approximately 7:00 PM CDT, Starship SN5 completed the 150 meter flight test, landing at an adjacent landing site, thus becoming the first full-scale prototype to perform a successful flight test. After the flight, Elon Musk stated that SpaceX would carry out several additional short flights in order to refine the launch process, before attempting a high altitude flight using a vehicle with body flaps.

Starship SN6 completed construction, and was moved to the test stand; static fire testing was to start on 23 August 2020, after 3 attempts on August 23rd, 2020 they finally had a successful static fire at about 7:43 PM CDT. After unsuccessful hop attempts on August 30th 2020, SN6 completed a maiden 150 meter flight test on September 3rd at about 12.45 PM CDT. Currently, other prototypes are under construction for pressure testing (Starship SN7.1), and flight test articles (Starship SN8 and SN9, the first to have nosecones and aero surfaces).

While the Starship program had a small development team for several years, and a larger development and build team since late 2018, Musk declared in June 2020 that Starship is the top SpaceX priority, except for anything related to reduction of Crew Dragon return risk for the NASA Demo-2 flight to the ISS, and remained so in September 2020.

While Mars has always been a critically important driver of requirements for the Starship system, and Musk had earlier hoped to have an initial orbital test flight by 2020, Musk toned down the expectations when he spoke to the Humans to Mars conference in August 2020. He said that that "the first Starship launches to orbit 'might not work,' saying that SpaceX is in 'uncharted territory. ...Nobody has ever made a fully reusable, orbital rocket'".

Musk clarified that SpaceX intends to fly exclusively cargo transport missions initially, and that passenger flights would come only much later. Starship will fly "hundreds of missions with satellites before we put people on board."

Types of Spaceships: Past, Present, and Future

5.2 USA-Moon Express MX-1E

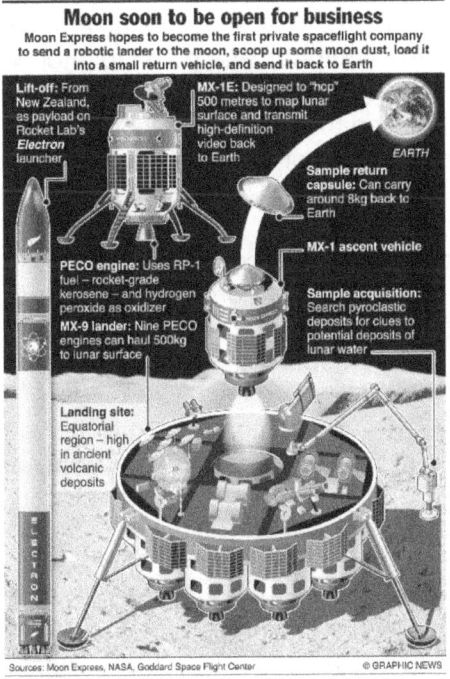

Moon soon to be open for business
Moon Express hopes to become the first private spaceflight company to send a robotic lander to the moon, scoop up some moon dust, load it into a small return vehicle, and send it back to Earth

Lift-off: From New Zealand, as payload on Rocket Lab's *Electron* launcher

MX-1E: Designed to "hop" 500 metres to map lunar surface and transmit high-definition video back to Earth

EARTH

Sample return capsule: Can carry around 8kg back to Earth

MX-1 ascent vehicle

PECO engine: Uses RP-1 fuel – rocket-grade kerosene – and hydrogen peroxide as oxidizer

MX-9 lander: Nine PECO engines can haul 500kg to lunar surface

Sample acquisition: Search pyroclastic deposits for clues to potential deposits of lunar water

Landing site: Equatorial region – high in ancient volcanic deposits

Sources: Moon Express, NASA, Goddard Space Flight Center © GRAPHIC NEWS

Moon Express (MoonEx; vehicle model prefix: MX) is an American privately held early-stage company formed by a group of Silicon Valley and space entrepreneurs. It had the goal of winning the $30 million Google Lunar X Prize, and of ultimately mining the Moon for natural resources of economic value. The company was not able to make a launch attempt to reach the Moon by March 31, 2018, the deadline for the prize. As of February 2020, Moon Express is focused on supporting NASA under its Commercial Lunar Payload Services (CLPS) contract.

Types of Spaceships: Past, Present, and Future

History

In August 2010, Robert D. Richards, Naveen Jain, and Barney Pell co-founded Moon Express, a Mountain View, California-based company that plans to offer commercial lunar robotic transportation and data services with a long-term goal of mining the Moon for resources, including elements that are rare on Earth, including niobium, yttrium and dysprosium.

Beginning in 2010, Moon Express based itself at the NASA Ames Research Center. Moon Express and NASA signed a contract in October 2010 for data purchase that could be worth up to US$10,000,000.

On June 30, 2011, the company held its first successful test flight of a prototype lunar lander system called the Lander Test Vehicle (LTV) that was developed in partnership with NASA. On September 11, 2011, Moon Express set up a robotics lab for a lunar probe named the "Moon Express Robotics Lab for INnovation" (MERLIN) and hired several engineering students who had successfully competed at the FIRST Robotics Competition.

In mid-2012, Moon Express started work with the International Lunar Observatory Association (ILOA) to put a shoebox-sized astronomical telescope called International Lunar Observatory on the Moon.

By 2012, MoonEx had 20 employees, and in December 2012, MoonEx acquired one of the other Google Lunar X-Prize teams, Rocket City Space Pioneers, from Dynetics for an undisclosed sum. The agreement made Tim Pickens, the former lead of the RCSP team, the Chief Propulsion Engineer for MoonEx.In September 2013,

Types of Spaceships: Past, Present, and Future

MoonEx added Paul Spudis as Chief Scientist and Jack Burns as Science Advisory Board Chair.

In October and November 2013, Moon Express conducted several free flight tests of its flight software utilizing the NASA Mighty Eagle lander test vehicle, under a Reimbursable Space Act Agreement with the NASA Marshall Space Flight Center. One month later, in December 2013, MoonEx unveiled the MX-1 lunar lander, a toroidal robotic lander that uses high-test hydrogen peroxide as its rocket propellant to support vertical landing on the lunar surface. On April 30, 2014 NASA announced that Moon Express was one of the three companies selected for the Lunar CATALYST initiative.

By December 2014, Moon Express successfully conducted flight tests of its "MTV-1X" lander test vehicle at the Kennedy Space Center Shuttle Landing Facility, becoming the first private company (and GLXP team) to demonstrate a commercial lunar lander test.

In 2015, the company announced that it would relocate to Florida's Cape Canaveral in 2015. In July 2016, Moon Express stated it would be taking over Cape Canaveral Launch Complexes 17 and 18.

On July 20, 2016, the Federal Aviation Administration approved Moon Express plans for a mission to deliver commercial payloads to the Moon, making Moon Express the first private company to receive government approval for a commercial space mission beyond traditional Earth orbit under the requirements of the Outer Space Treaty.

On October 31, 2017, NASA extended the agreement for the Lunar CATALYST initiative for 2 more years.

Types of Spaceships: Past, Present, and Future

On July 12, 2018, both historic launch towers at Space Launch Complex 17 were demolished via controlled demolition to make way for Moon Express facilities to test its lunar lander. That month, Moon Express was unable to make payroll and laid off nine employees; the employees did not receive back-pay until October 2018.

In October 2018, the company signed several collaboration agreements with the Canadian Space Agency (CSA) and a number of Canadian aerospace companies.

On November 29, 2018, Moon Express joined the Commercial Lunar Payload Services program of NASA, becoming eligible to bid on delivering science and technology payloads to the Moon for NASA.

5.3 United Kingdom-Reaction Engines Limited Skylon

Skylon is a series of designs for a single-stage-to-orbit spaceplane by the British company Reaction Engines Limited (REL), using SABRE, a combined-cycle, air-breathing rocket propulsion system. The vehicle design is for a hydrogen-fueled aircraft that would take off from a purpose-built runway, and accelerate to Mach 5.4 at 26 kilometers' (85,000 ft) altitude (compared to typical airliners' 9–13 kilometers' or 30,000–40,000 feet) using the atmosphere's oxygen before switching the engines to use the internal liquid oxygen (LOX) supply to take it into orbit. It could carry 17 tons (37,000 lb) of cargo to an equatorial low Earth orbit (LEO); up to 11 tons (24,000 lb) to the International Space Station, almost 45% more than the capacity of the European Space Agency's Automated Transfer Vehicle; or 7.3 tons; 7,300 kilograms (16,000 lb) to Geosynchronous Transfer Orbit (GTO), over 24% more than SpaceX Falcon 9 launch vehicle in reusable mode (As of 2018.)

The relatively light vehicle would then re-enter the atmosphere and land on a runway, being protected from the conditions of re-entry by a ceramic composite skin. When on the ground, it would undergo inspection and necessary maintenance, with a turnaround time of approximately two days, and be able to complete at least 200 orbital flights per vehicle.

As work on the project has progressed, information has been published on a number of design versions, including A4,Fig 1 p165 C1, C2,and D1. Testing of the key technologies was successfully completed in November 2012, allowing Skylon's design to advance from its research phase to a development phase. As of 2017, an engine test facility was being built at Westcott and if all goes to plan, the first ground-based engine tests could happen in 2020, and SABRE engines could be performing uncrewed test flights by 2025.

In paper studies, the cost per kilogram of payload carried to LEO in this way is hoped to be reduced from the current £1,108/kg (as of December 2015), including research and development, to around £650/kg, with costs expected to fall much more over time after initial expenditures have amortised. In 2004, the developer estimated the total lifetime cost of the Skylon C1 program to be about $12 billion. As of 2017, only a small portion of the funding required to develop and build Skylon had been secured. For the first couple of decades the work was privately funded, with public funding beginning in 2009 through a European Space Agency (ESA) contract. The British government pledged £60 million to the project on 16 July 2013 to allow a prototype of the SABRE engine to be built, contracts for this funding were signed in 2015.

5.4 Next Generation Chinese Spacecraft

Next-generation crewed spacecraft is a type of reusable spacecraft developed and manufactured by China Aerospace Science and Technology Corporation (CASC). The prototype of the spacecraft underwent its first uncrewed test flight on 5 May 2020.

The crew carrier is designed to ferry astronauts to the Chinese space station in Earth orbit as well as conducting lunar exploration.

Intended to replace the Shenzhou spacecraft, the new vehicle is larger and lunar-capable. It consists of two modules: a crew module that returns to Earth, and an expendable service module to provide propulsion, power and life support for the crew module while in space. It is capable of carrying six astronauts, or three astronauts and 500 kg of cargo. The new crew module is partially reusable with its detachable heat shields, while the spacecraft as a whole features a modular design that allows it to be constructed to meet different mission demands. With its propulsion and power module, the crew spacecraft measures nearly 8.8 meters long. It weighs around 21600 kg fully loaded with equipment and propellant, according to the China Manned Space Engineering Office (CMSEO). Lunar missions are expected in the 2030s.

5.5 India-Gaganyaan

Gaganyaan (Sanskrit; IAST: gagan-yāna) transl. "Sky Craft") is an Indian crewed orbital spacecraft intended to be the formative spacecraft of the Indian Human Spaceflight Program. The spacecraft is being designed to carry three people, and a planned upgraded version will be equipped with rendezvous and docking capability. In its maiden crewed mission, Indian Space Research Organization (ISRO)'s largely autonomous 3.7-tonne (8,200 lb) capsule will orbit the Earth at 400 km (250 mi) altitude for up to seven days with a two or three-person crew on board. The crewed vehicle is planned to be launched on ISRO's GSLV Mk III in December 2021. This Hindustan Aeronautics Limited (HAL) manufactured crew module had its first un-crewed experimental flight on 18 December 2014. As of May 2019, design of the crew module has been completed. Defense Research and Development Organization (DRDO) will provide support for critical human-centric systems and technologies like space grade food, crew healthcare, radiation measurement and protection, parachutes for the safe recovery of the crew module and fire suppression system.

Types of Spaceships: Past, Present, and Future

On 11 June 2020, it was announced that while the first uncrewed Gaganyaan launch has been delayed due to COVID-19 pandemic in India, overall timeline for crewed launches is expected to remain unaffected.

Funding and infrastructure

A crewed spacecraft would require about ☐ 124 billion (US$1.77 billion) over a period of seven years, including the ☐ 50 billion (US$0.7 billion) for the initial work of the crewed spacecraft during the Eleventh Five-Year Plan (2007–12) out of which govt released ☐ 500 million (US$7 million) in 2007-08. In December 2018, the government approved further ☐ 100 billion (US$1.5 billion) for a 7-days crewed flight of 3 astronauts to take place by 2021.

Madhavan Chandradathan, director of Satish Dhawan Space Centre (SDSC), stated that ISRO would need to set up an astronaut training facility in Bangalore. Newly established Human Space Flight Centre (HSFC) will coordinate the IHSF efforts. Existing launch facilities will be upgraded for launches under Indian Human Spaceflight project with extra facilities needed for launch escape systems. Russia is likely to provide astronaut training, and assist with some aspects in the development of the launcher. In spring 2009 the full-scale mock-up of crew capsule of Gaganyaan was built and delivered to Satish Dhawan Space Centre for training of astronauts.

India has already successfully developed and tested several building blocks, including re-entry space capsule, pad abort test, safe crew ejection mechanism in case of rocket failure, flight suit developed by DEBEL and the powerful GSLV-MkIII launch vehicle. Having met all required technological keystones, the Indian Human Spaceflight Program was accepted and formally announced by the Prime Minister Narendra Modi on 15 August 2018. Gaganyaan will be the first crewed spacecraft under this program.

ISRO's Human Space Flight Centre and Glavcosmos, which is a subsidiary of the Russian state corporation Roscosmos, signed an agreement on July 1, 2019 for cooperation in the selection, support, medical examination and space training of Indian astronauts. An ISRO Technical Liaison Unit (ITLU) will be setup in Moscow to facilitate the development of some key technologies and establishment of special facilities which are essential to support life in space.

On 25 October 2019, ISRO's Human Space Flight Centre and Glavcosmos signed a contract to evaluate the possibility of using Russian life support systems and thermal control for Gaganyaan.

Types of Spaceships: Past, Present, and Future

5.6 Russia-Orel

Orel (Russian: Орёл, lit. 'Eagle') or Oryol, formerly Federation is a project by Roscosmos to develop a new-generation, partially reusable crewed spacecraft.

Until 2016, the official name was (New Generation Piloted Transport Ship) or PTK NP. The goal of the project is to develop a next-generation spacecraft to replace the Soyuz spacecraft developed by the former Soviet Union to support low Earth orbit and lunar operations. It is similar in function to the US Orion or Commercial Crew Development spacecraft.

The PPTS project was started following a failed attempt by Russia and the European Space Agency (ESA) to co-develop the Crew Space Transportation System (CSTS). Following ESA member states declining to finance Kliper in 2006 over concerns about workshare then again declining to finance development of CSTS in 2009 over technology

Types of Spaceships: Past, Present, and Future

transfer to Russia that could be used for military purposes, the Russian Federal Space Agency ordered a new crewed spacecraft from Russian companies. A development contract was awarded to RKK Energia on 19 December 2013.

Orel is intended to be capable of carrying crews of four into Earth orbit and beyond on missions of up to 30 days. If docked with a space station, it could stay in space up to a year, which is double the duration of the Soyuz spacecraft. The spacecraft will send cosmonauts to lunar orbit, with a plan to place a space station there, called Lunar Orbital Station

Previously, ESA officials had inquired whether they could be part of the Constellation Program of the United States, with NASA focused on its Orion spacecraft, but they had received a negative response. Consequently, Europe decided to join the Russians to co-develop a new-generation crewed spacecraft. ESA insisted on a joint design rather than the Russian-designed Kliper, and as a result the joint Russian/European CSTS project came into being. CSTS had completed an initial study phase, which lasted for 18 months from September 2006 to spring 2008, before the project was shut down before an ESA member state conference in November 2008. ESA decided to use some technology of the CSTS project from its Automated Transfer Vehicle.

The Russian space agency, Roskosmos, had repeatedly received proposals from Moscow-based Khrunichev enterprise to develop a new-generation crewed spacecraft based on the TKS spacecraft that would be launched on the new Angara launch Vehicle. Citing the requirement to start work on a new crewed spacecraft, Russia decided to go forward with the project by itself.

Types of Spaceships: Past, Present, and Future

By the first quarter of 2009, Roskosmos had finalized its requirements for the next-generation crewed spacecraft and had received proposals from both RKK Energia and Khrunichev enterprise. This was the actual beginning of the PPTS project. The agency was finally ready to name the prime developer of the vehicle. Formally, only two organizations which were practically capable of developing crewed space vehicles competed in the government tender to build the new spacecraft—RKK Energia in Korolev and Moscow-based Khrunichev enterprise.

Although Roskosmos has remained tight-lipped about the project, a number of Russian officials made statements hinting about various stages of the project. On 21 January 2009, the head of Roskosmos, Anatoly Perminov, told Rossiyskaya Gazeta, a Russian newspaper, that Russia would likely proceed with independent development of the next-generation crewed spacecraft. According to Perminov, the agency and its main research and certification center—TsNIIMash—had already conducted an expanded meeting of the Scientific and Technical Council, NTS, examining follow-on transport systems, including the next-generation crewed ship. It would be followed by a government tender to select a developer for the new vehicle. Perminov further indicated that the new spacecraft would be expected to enter service within a time frame of the Orion spacecraft, but a more detailed development plan would be ready with the preliminary design of the vehicle in the middle of 2010.

In the first quarter of 2009, Roskosmos released requirements which were used in the development of the Technical Assignment to the industry working on the PPTS project. The preliminary development of the project was expected to take place from March 2009 until June 2010 at an estimated cost of around 800,000,000 rubles ($24 million). The work apparently covered only an Earth-orbiting version of the spacecraft, while laying the foundation for later lunar-orbiting spacecraft, or even a Mars-bound crew vehicle.

The agency's general requirements asked the industry to develop a vehicle of "foreign" standards in its technical capabilities and cost, while at the same time using existing technologies as much as possible.

In November 2019, it was announced that the first test flight was scheduled for 2023, and the first crewed flight for 2025.

Types of Spaceships: Past, Present, and Future

6.0 Spaceship Concepts

While there are many science fiction concepts for exploring the Solar System, I've only found one—the NASA Deep Space Transport proposed by a major space agency or company. (Besides the Space X Starship project)

6.1 USA-NASA Deep Space Transport

The Deep Space Transport (DST), also called Mars Transit Vehicle, is a crewed interplanetary spacecraft concept by NASA to support science exploration missions to Mars of up to 1,000 days. It would be composed of two elements: an Orion capsule and a propelled habitation module. As of late 2019, the DST is still a concept to be studied, and NASA has not officially proposed the project in an annual U.S. federal government budget cycle. The DST vehicle would depart and return from the Lunar Gateway to be serviced and reused for a new Mars mission.

Types of Spaceships: Past, Present, and Future

Architecture overview

Both the Gateway and the DST would be fitted with the International Docking System Standard. The DST spacecraft would comprise two elements: an Orion capsule and a habitation module that would be propelled by both electric propulsion and chemical propulsion, and carry a crew of four in a medium-sized habitat. The fully assembled spacecraft with the Orion capsule mated, would have a mass of about 100 metric tons. The spacecraft's habitat portion will likely be fabricated using tooling and structures developed for the SLS propellant tank; it would be 8.4 m (28 ft) in diameter and 11.7 m (38 ft) in length.

The habitat portion of the DST spacecraft may also be equipped with a laboratory with research instrumentation for physical sciences, electron microscopy, chemical analyses, freezers, medical research, small live animal quarters, plant growth chambers, and 3D printing. External payloads might include cameras, telescopes, detectors, and a robotic arm.

Its initial target for exploration is Mars (flyby or orbit), and other suggested destinations are Venus (flyby or orbit), and a sample return from a large asteroid. If the DST spacecraft were to orbit Mars, it would enable opportunities for real-time remote operation of equipment on the Martian surface, such as a human-assisted Mars sample return.

It would use a lunar flyby to build up speed and then using solar electric propulsion (SEP) it would accelerate into a heliocentric orbit. There it would complete its transit to Mars or other possible destinations. It would use chemical propulsion to enter Mars orbit. Crews could perform remote observations or depart for the surface during a 438-day window. The vehicle would depart Mars orbit via a chemical burn. It would use a mix of SEP and lunar gravity assists to recapture into Earth's sphere of influence.

Types of Spaceships: Past, Present, and Future

7.0 Summary

There are many more past spaceship designs which have been built for human flight than I thought when I began this project.

There are also a lot more spaceships being built than I had any idea for usage in the decades to come.

The creativity of the human spirit is on display in these projects and the future of the human exploration of our solar system is bright. I just hope I live to see many of these incredible journeys come to fruition.

Martin K. Ettington

September 2020

Types of Spaceships: Past, Present, and Future

Types of Spaceships: Past, Present, and Future

8.0 Bibliography

1. Top Private Spaceships. *https://www.space.com/15735-top-private-spaceships-countdown.html.* [Online]

2. Proposed Spacecraft. *https://en.wikipedia.org/wiki/Category:Proposed_spacecraft.* [Online]

3. List of Crewed Spacecraft. *https://en.wikipedia.org/wiki/List_of_crewed_spacecraft.* [Online]

4. North Amaerican X-15. *https://en.wikipedia.org/wiki/North_American_X-15.* [Online]

5. Vostok Spacecraft. *https://en.wikipedia.org/wiki/Vostok_(spacecraft).* [Online]

6. Project Mercury. *https://en.wikipedia.org/wiki/Project_Mercury.* [Online]

7. Project Gemini . *https://en.wikipedia.org/wiki/Project_Gemini.* [Online]

8. Soviet Voskhod Spacecraft. *https://en.wikipedia.org/wiki/Voskhod_(spacecraft).* [Online]

9. Apollo Command Module. *https://en.wikipedia.org/wiki/Apollo_command_and_service_module.* [Online]

10. Space Shuttle Orbiter. *https://en.wikipedia.org/wiki/Space_Shuttle_orbiter.* [Online]

Types of Spaceships: Past, Present, and Future

11. Soviet-Buran Shuttle.
https://en.wikipedia.org/wiki/Buran_(spacecraft). [Online]

12. Soyuz Spacecraft.
https://en.wikipedia.org/wiki/Soyuz_(spacecraft). [Online]

13. SpaceX Dragon 2.
https://en.wikipedia.org/wiki/SpaceX_Dragon_2. [Online]

14. Boeing Starliner.
https://en.wikipedia.org/wiki/Boeing_Starliner. [Online]

15. Orion Spacecraft.
https://en.wikipedia.org/wiki/Orion_(spacecraft). [Online]

16. Space Ship Two.
https://en.wikipedia.org/wiki/SpaceShipTwo. [Online]

17. Dream Chaser-Crewed Version.
https://en.wikipedia.org/wiki/Dream_Chaser#Crewed_version. [Online]

18. Space X Starship.
https://en.wikipedia.org/wiki/SpaceX_Starship. [Online]

19. Moon Express.
https://en.wikipedia.org/wiki/Moon_Express. [Online]

20. Skylon Spacecraft.
https://en.wikipedia.org/wiki/Skylon_(spacecraft). [Online]

21. Chinese Next Generation Crewed Spacecraft.
https://en.wikipedia.org/wiki/Next-generation_crewed_spacecraft. [Online]

22. Russian Orel Spacecraft.
https://en.wikipedia.org/wiki/Orel_(spacecraft). [Online]

23. Russian Orel Spacecraft.
https://en.wikipedia.org/wiki/Orel_(spacecraft). [Online]

24. Deep Space Transport.
https://en.wikipedia.org/wiki/Deep_Space_Transport.
[Online]

Types of Spaceships: Past, Present, and Future

Types of Spaceships: Past, Present, and Future

9.0 Index

Types of Spaceships: Past, Present, and Future